I0606185

Nature Is a Sculptor

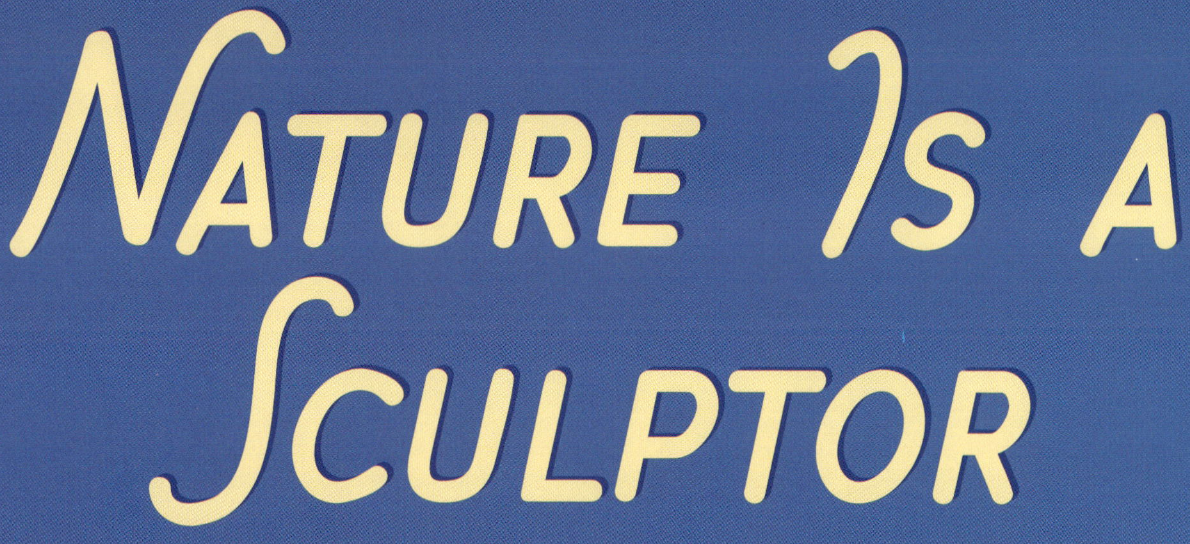

Weathering and Erosion

Heather Ferranti Kinser

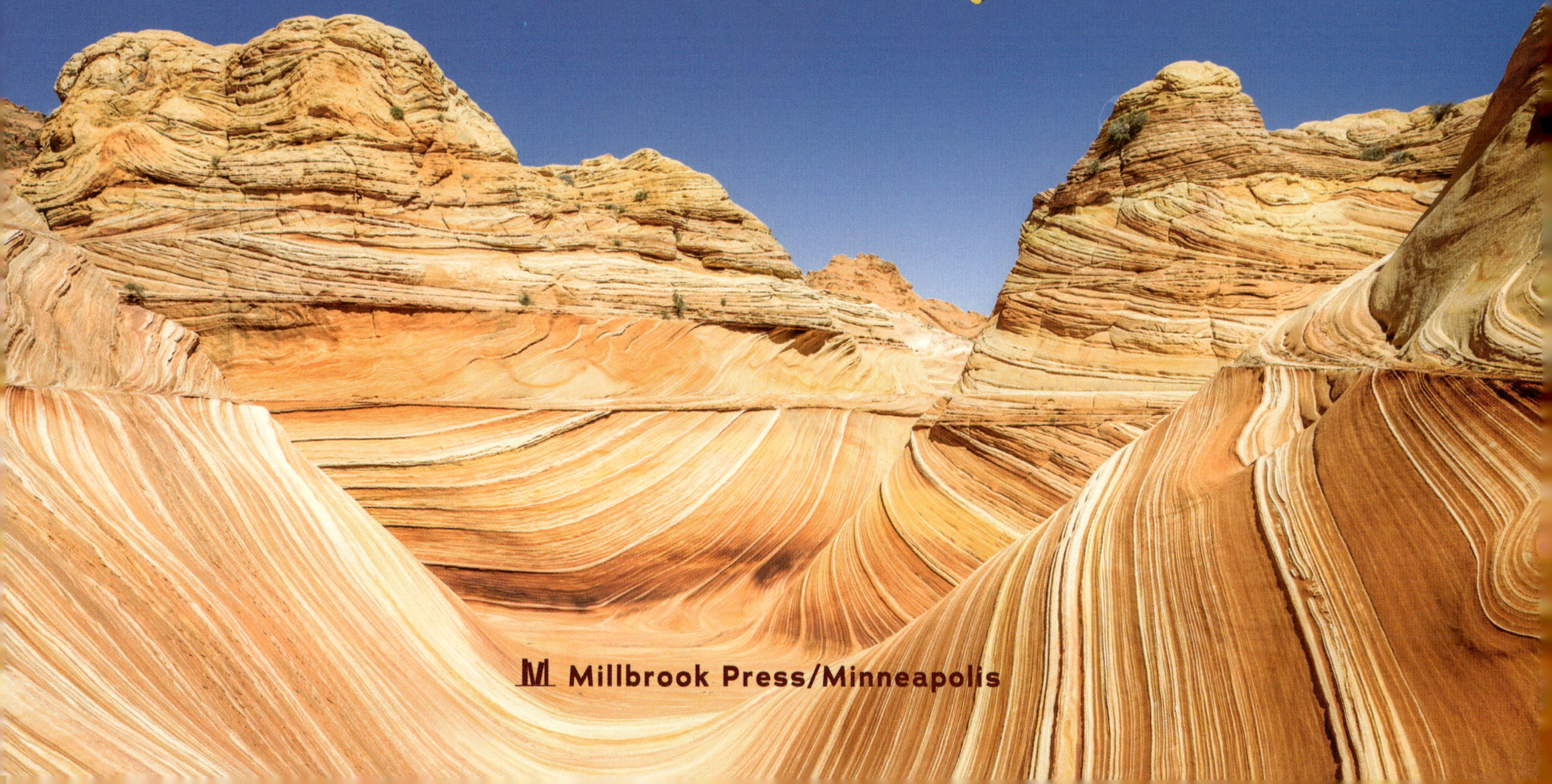

Millbrook Press/Minneapolis

TARA RIVER, DURMITOR NATIONAL PARK, MONTENEGRO

GREWINGK GLACIER, ALASKA
ice,
and wind . . .
CORAL PINK SAND DUNES STATE PARK, UTAH

it **etches**, **scrapes**, and **carves** the face of **towers**,

SPIDER ROCK, CANYON DE CHELLY, ARIZONA

canyons,

SANTA ELENA CANYON, BIG BEND NATIONAL PARK, TEXAS

cliffs.

CABRILLO NATIONAL MONUMENT, CALIFORNIA

It **shapes** and **shaves**

a **dome**,

an **arch**,

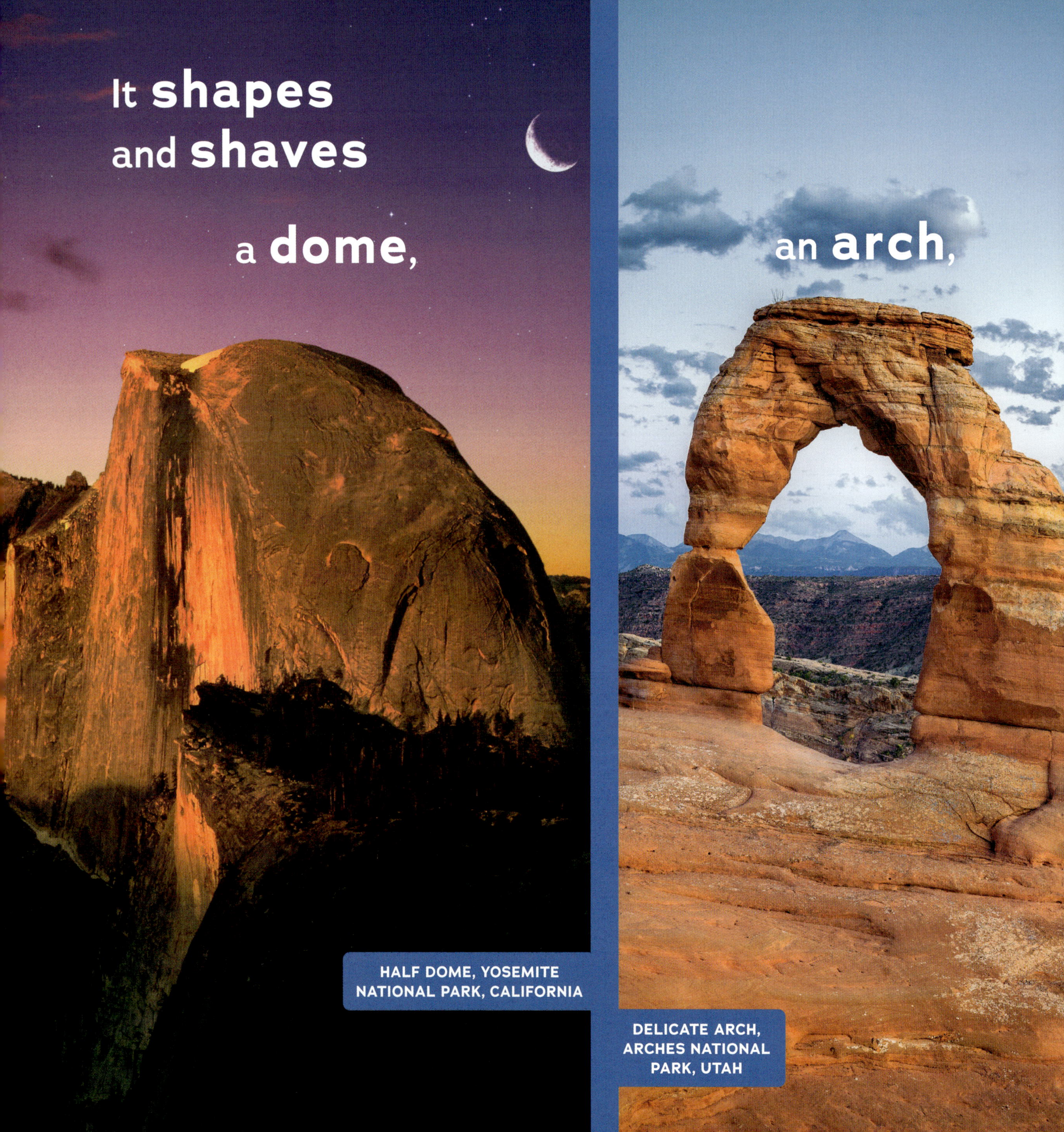

HALF DOME, YOSEMITE NATIONAL PARK, CALIFORNIA

DELICATE ARCH, ARCHES NATIONAL PARK, UTAH

a hoodoo,
GOBLIN VALLEY STATE PARK, UTAH
or a wave.
THE WAVE, VERMILLION CLIFFS NATIONAL MONUMENT, ARIZONA

It decorates the landscape,

slowly

BASALT COLUMNS, DEVILS POSTPILE NATIONAL MONUMENT, CALIFORNIA

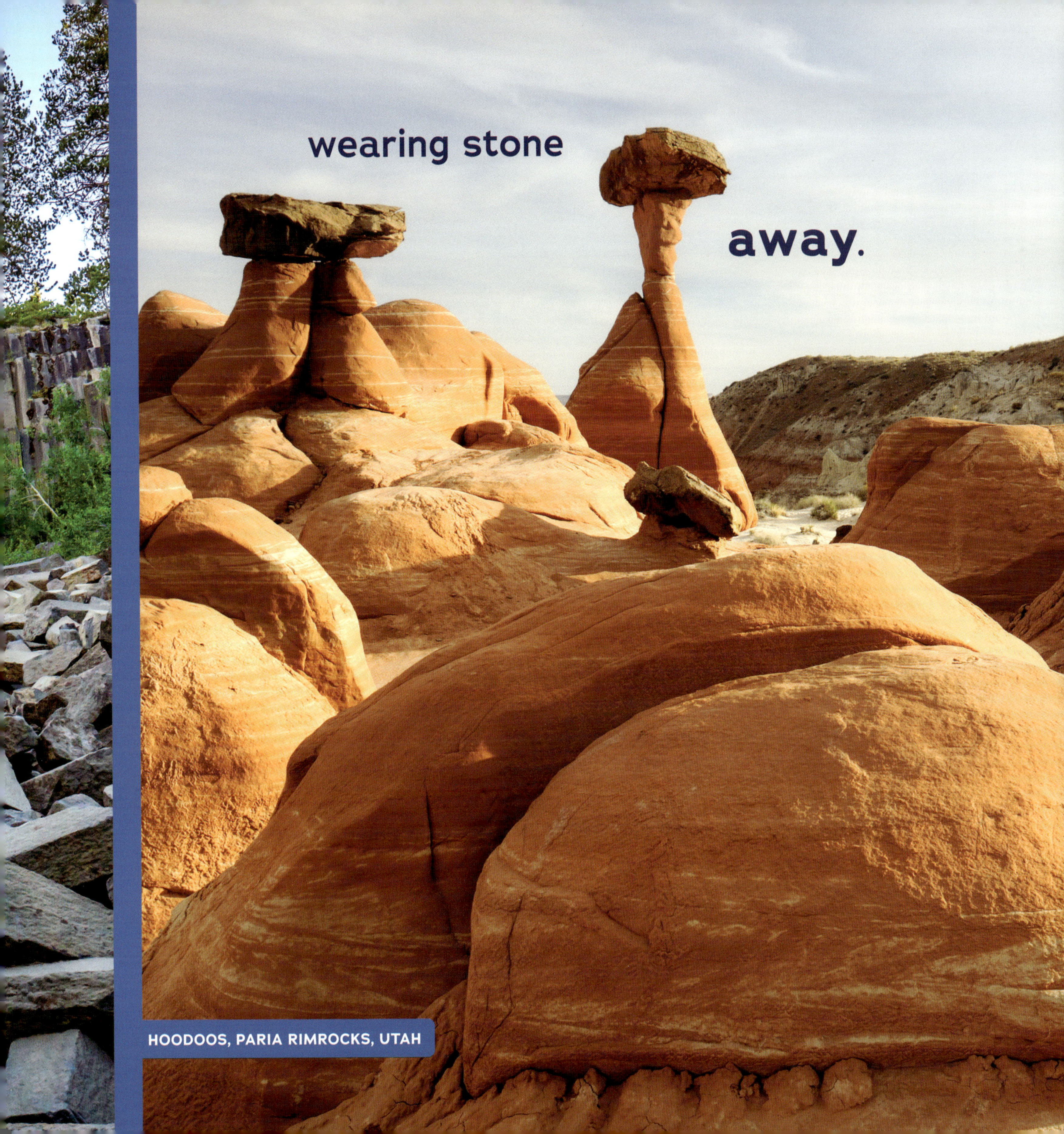

wearing stone

away.

HOODOOS, PARIA RIMROCKS, UTAH

Nature is a sculptor.

It **weathers**

ANTELOPE CANYON, ARIZONA

and **erodes**.

Over time, the tallest peaks

are **whittled**,

chipped,

GRAND TETON NATIONAL PARK, WYOMING

and **blown**.
CAPE KIWANDA STATE NATURAL AREA, OREGON

The ocean is a **hammer**

pounding shorelines into bits.

PEGGY'S COVE, NOVA SCOTIA, CANADA

Ice—a chilly chisel—finds a crack,

expands,

and **splits**.

SPLIT APPLE ROCK, ABEL TASMAN NATIONAL PARK, KAITERITERI, NEW ZEALAND

Ice can also be a **grinder**

when a glacier's
on the move.

RUSSELL GLACIER, GREENLAND ICE SHEET, KANGERLUSSUAQ, GREENLAND

ANZA-BORREGO DESERT STATE PARK, CALIFORNIA

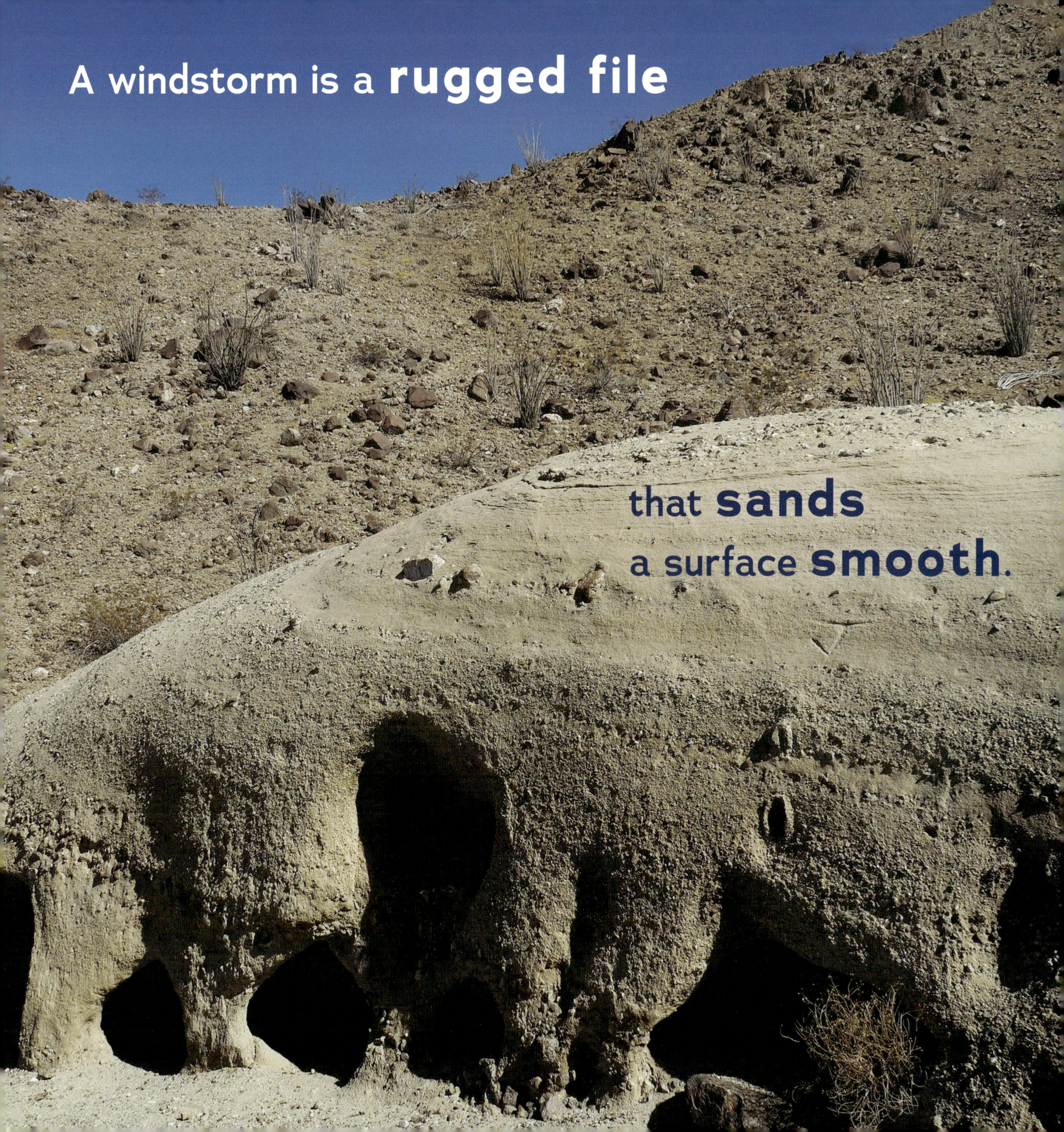

A windstorm is a **rugged file**

that **sands**
a surface **smooth**.

Rivers **gouge**.

KAWARAU GORGE, OTAGO, NEW ZEALAND

Waters **dissolve**,

opening a cave.

Cold nights after hot, dry days will **flake** and **break** and **rearrange**.

LURAY CAVERNS, VIRGINIA

MOUNT KINABALU, SABAH, MALAYSIA

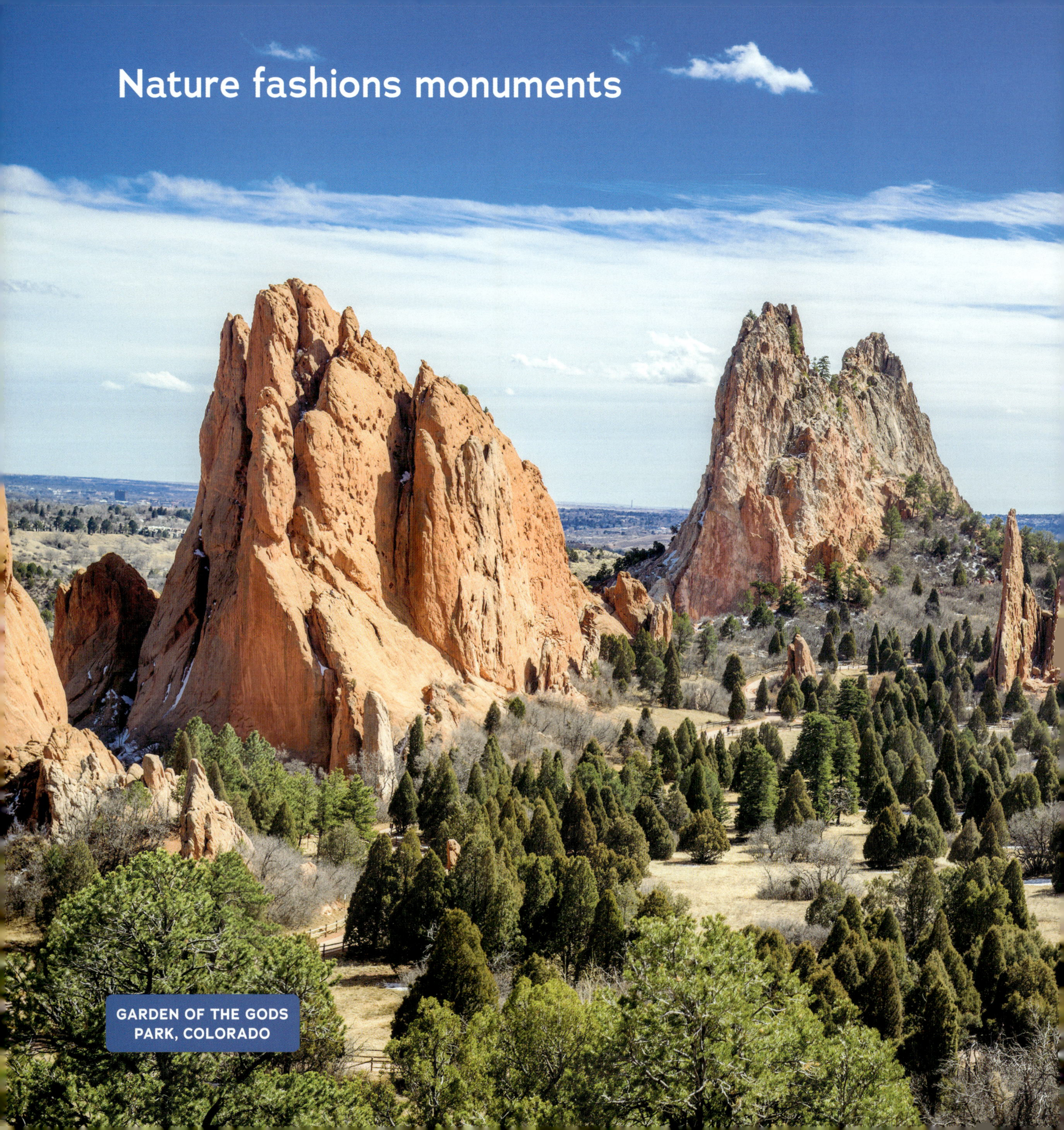

Nature fashions monuments

GARDEN OF THE GODS PARK, COLORADO

and marvels
big

DEVILS TOWER NATIONAL MONUMENT, WYOMING

and **small**.

TAFONI ROCK FORMATIONS, BEAN HOLLOW STATE BEACH, CALIFORNIA

Some go on **forever** . . .

some are **barely there** at all.

GRAND CANYON NATIONAL PARK, ARIZONA

BLACK SAND, PUNALU'U BEACH, HAWAII

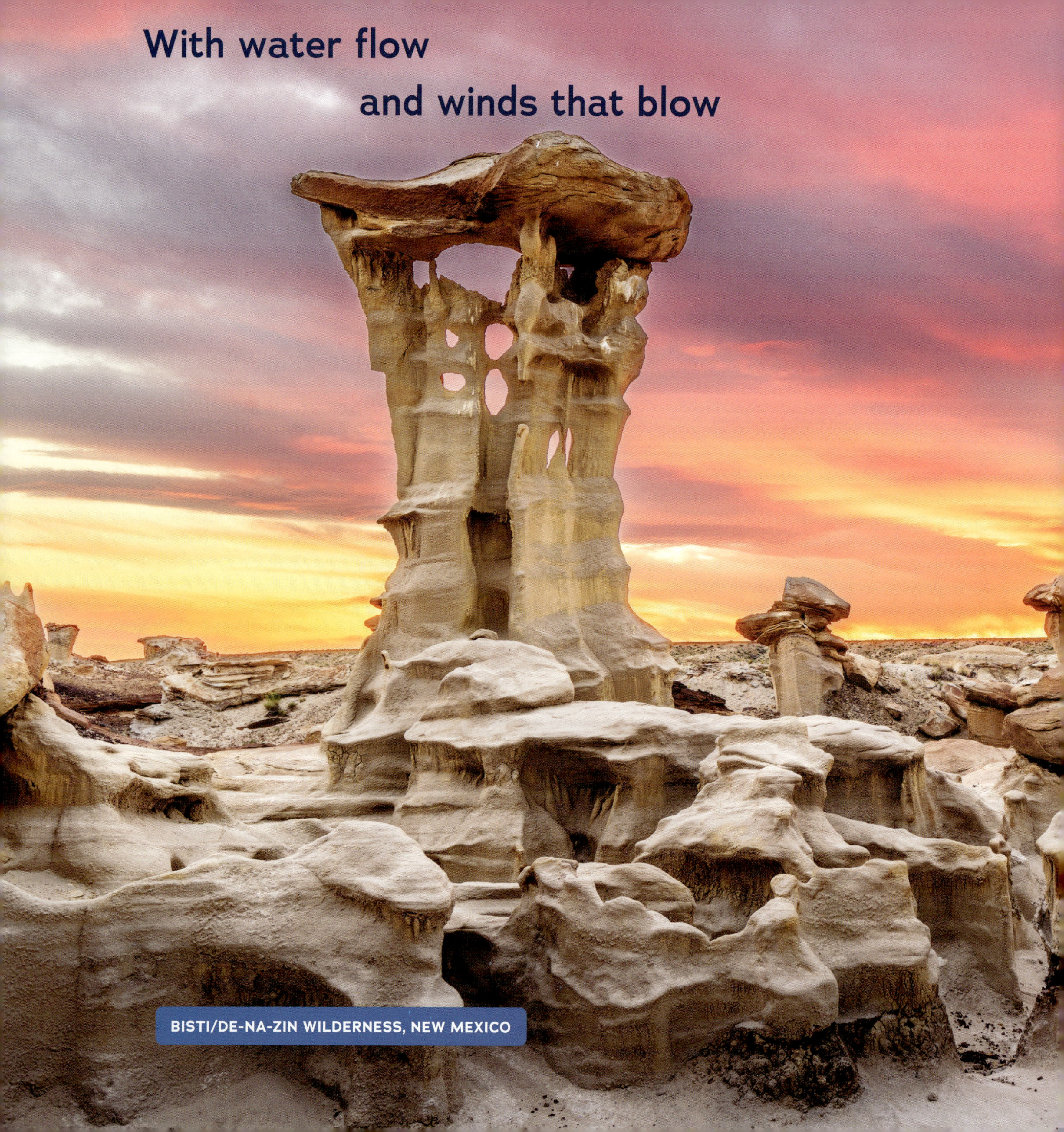
With water flow
and winds that blow
BISTI/DE-NA-ZIN WILDERNESS, NEW MEXICO

and icy-cold extremes,

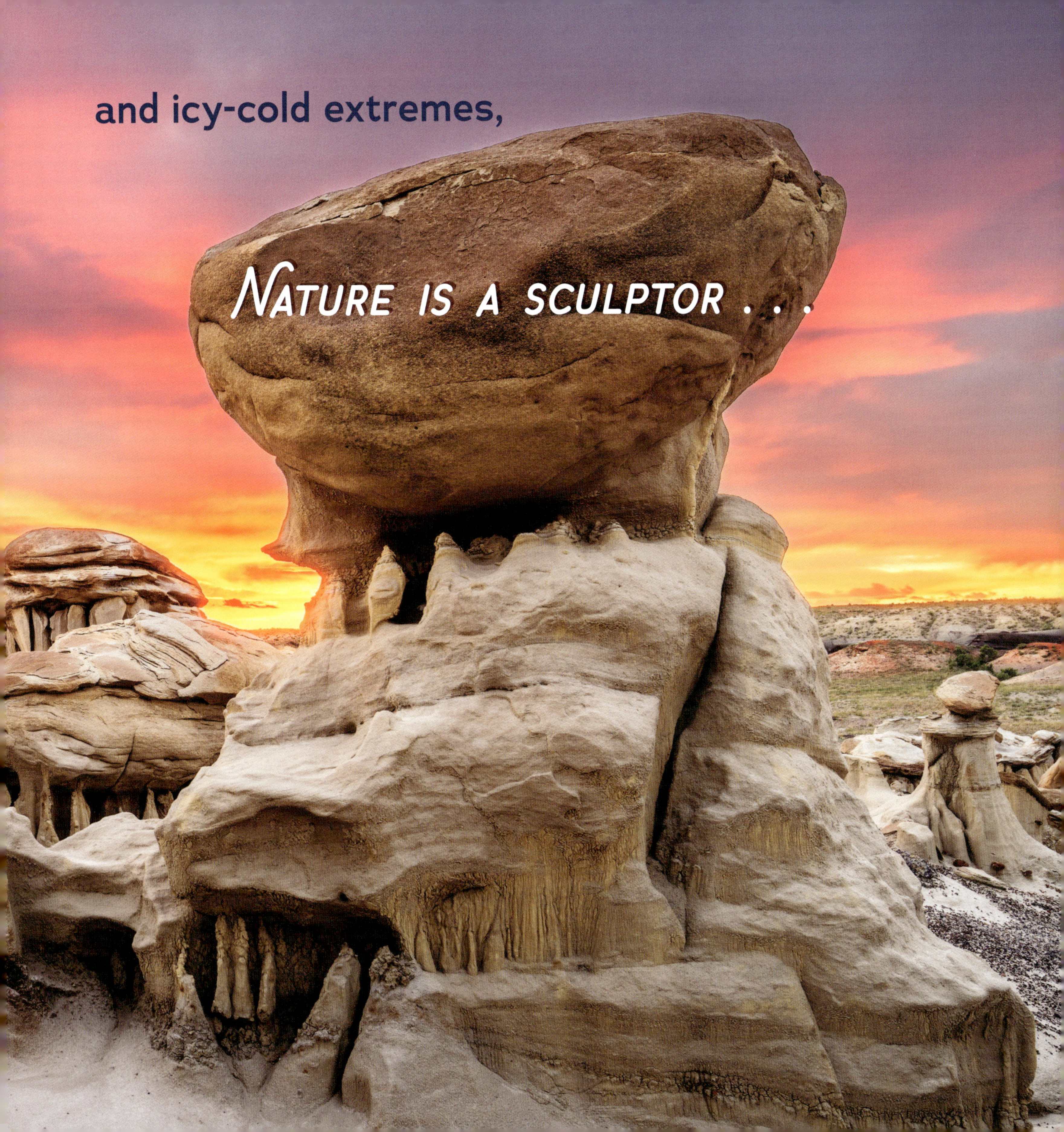

Nature is a sculptor . . .

and the world

BRYCE CANYON NATIONAL PARK, UTAH

is our museum.

About Nature's Sculptures

Weird and wonderful rock formations are nature's unique creations. But how does nature actually "make" them? Through processes called *weathering*, *erosion*, and *deposition*. Together, these processes create amazing natural sculptures.

Weathering

Weathering occurs when rock or soil breaks down or dissolves. Soft types of rock, such as sandstone or limestone, weather faster. Harder rocks, such as granite, weather slowly. There are two kinds of weathering: chemical and mechanical. *Chemical weathering* occurs when rocks are broken down by a chemical reaction. For instance, the most common type of cave system forms when acids in water dissolve soft rock into tiny particles and then wash the particles away. *Mechanical weathering* occurs when rocks are changed by a physical force. For example, when water freezes in a crack in a rock, the expanding ice takes up more space and can break the rock apart.

Erosion

Erosion happens when rocks or soil wear away and are transported, or moved, to another place. Erosion can occur slowly or quickly. A river-carved canyon is an example of erosion that has taken place slowly, over thousands of years. As water flows over the ground, it erodes rock and soil and transports the debris downstream. Each year, the river carves the canyon deeper and deeper. A landslide, on the other hand, causes erosion to happen suddenly. When rock and soil have already been broken down by weathering, it is easier for the forces of erosion to move them. One of the main forces of erosion is gravity.

Deposition

The opposite of erosion is *deposition*, the process of building up a new landform out of eroded materials. Sand dunes are an example of deposition, since they are formed when winds move sand from one place to another, creating a large pile. Glaciers, which are enormous rivers of slowly moving ice, can also deposit rocks and soil in new locations.

Nature's Sculpting Tools

Water

Water has a powerful effect on rock and soil. Rivers chip and tumble rock. Over time they can cut gorges and canyons into the ground. Heavy rain can cause landslides. Ocean waves pound rock into smaller and smaller bits until it becomes sand. Rainwater can seep into soft stone and gradually carve a mazelike cave system.

Ice

How can ice break rock? Water slips into tiny fractures in the rock and then freezes. The frozen water expands, cracking the rock over time. Additionally, in Earth's coldest regions, icy glaciers shift and scrape any landforms that lie in their path, sometimes even picking up and moving boulders.

Wind

Wind can also cause erosion. Strong winds blow soil away, leaving behind bare rock. Around oceans or in deserts, wind picks up sand and dust. These particles blowing on the wind can act like sandpaper, scraping and smoothing rock surfaces. Wind can also shift and swirl sand into rolling, wavy dunes.

Other Tools

A few other natural causes of weathering and erosion are gravity and plant roots. Frequent changes between extreme heat and freezing cold can also cause rock to crumble and flake, since rock expands in the heat and shrinks in the cold. Over time, this process weakens the rock. People can trigger erosion too, such as when they move soil or rock during construction projects or farming.

Nature's Sculpture Gallery

Hoodoos

A hoodoo is a thin tower or "mushroom" of rock formed when soft and hard rock layers break down unevenly. A layer of hard rock on top of the hoodoo acts like a helmet, protecting the softer rock below it. Eventually, the hoodoo is weathered by rain or cracked by ice. The hard rock wears away slowly, while the softer rock weathers more quickly, leaving behind lumpy, bulging shapes.

Basalt Columns

Basalt columns begin as lava flows. When a particular kind of fast-flowing lava fills up a valley, it creates a lava lake that cools and hardens faster above than below. Like a mud puddle on a sunny day, the hardening lava cracks on the surface. When the lava cools completely, it becomes a type of rock called basalt. Over thousands of years, the rock cracks and eventually splits all the way down to the base of the valley, forming columns. From above, basalt columns look like six-sided puzzle pieces fitted closely together.

Canyons

Most canyons are carved by rivers over long stretches of time. A river wears down and carries away rock with the force of its flowing water and can cut even deeper when it floods. As the river cuts deeper, the canyon walls can crumble, making the canyon wider as well. Canyons come in many shapes and sizes. One type is called a *slot canyon*. These narrow, smooth-sided canyons form when floodwaters cut down through cracks in rock.

The Wave

This unusual feature is made of sandstone containing ancient sand dunes. Over time, rainwater flowed over it, forming grooves. The grooves were slowly weathered by desert winds into the smooth waves we see today.

Half Dome

This famous feature overlooking California's Yosemite Valley is the remains of a hardened magma chamber from a long-ago volcano. It is made of a hard rock called granite. During the last ice age, a glacier filled the river valley and scraped along its granite walls. The glacier created Half Dome, and even more chunks of granite have broken away over time as the dome weathered and eroded.

Tafoni

Tafoni are lacy "honeycomb" formations in rock. They form when a specific type of rock is repeatedly wetted and dried. This rock is porous, or covered in tiny holes called pores. Water enters the pores and weakens the rock, creating pits and pockets.

Glossary

acid: a chemical that can dissolve rocks or metals

chemical reaction: a chemical change that occurs when different substances mix

chemical weathering: a natural process in which rock is broken down or changed by chemical reactions

chisel: a stonecutting tool, made of strong metal, that can break rock when struck on one end with a hammer

deposit: something that piles up over time

force: something that causes objects to move and change

formation: a structure or deposit that takes on form and shape

fracture: a crack or break

glacier: an enormous, slowly moving river of ice

gravity: the force that causes things to fall

ice age: periods of time in Earth's history when glaciers covered much of the land

landslide: an event where a mass of rocks or soil suddenly tumbles downhill

mechanical weathering: a natural process in which rock is broken down or changed by physical forces

particle: a very small amount of something

Further Reading

Books

Aston, Dianna Hutts. *A Rock Is Lively.* San Francisco: Chronicle Books, 2015.

Chin, Jason. *Grand Canyon.* New York: Roaring Brook, 2017.

Pilutti, Deb. *Old Rock (Is Not Boring).* New York: G. P. Putnam's Sons, 2020.

Salas, Laura Purdie. *A Rock Can Be . . .* Minneapolis: Millbrook Press, 2015.

Websites

National Geographic: Erosion
https://education.nationalgeographic.org/resource/erosion
Check out this site to learn more about erosion.

National Geographic: Weathering
https://education.nationalgeographic.org/resource/weathering
Follow this link to learn more about weathering.

Sciencing: Weathering and Erosion
https://sciencing.com/difference-between-weathering-erosion-kids-8627014.html
Visit this page for answers to even more questions about weathering and erosion.

Photo Acknowledgments

Image credits: DCrane08/iStock/Getty Images, pp. 1, 29; mastermilmar/iStock/Getty Images, p. 2; Andre Pinto/Moment/Getty Images, p. 3 (top); Alfons Hauke/imageBROKER/Getty Images, p. 3 (bottom); Howard Grill/Moment/Getty Images, p. 4; Kanokwalee Pusitanun/iStock/Getty Images, pp. 5 (top), 29; Pateo/500px/Getty Images, p. 5 (bottom); Bill Ross/The Image Bank/Getty Images, pp. 6 (left), 29; Jakob Radlgruber/EyeEm/Getty Images, p. 6 (right); MJFelt/iStock/Getty Images, p. 7 (top); Shawn Walters/EyeEm/Getty Images, p. 7 (bottom); Png-Studio/iStock/Getty Images, pp. 8, 29; Rezus/iStock/Getty Images, pp. 9, 29; Left_Coast_Photographer/iStock/Getty Images, p. 10; Ron and Patty Thomas/E+/Getty Images, p. 12; Dale Boettcher/Moment/Getty Images, p. 13; Willowpix/E+/Getty Images, p. 14; jimfeng/iStock/Getty Images, p. 15; Jason Edwards/The Image Bank/Getty Images, p. 16; Federica Grassi/Moment/Getty Images, pp. 17, 21 (bottom), 29; travellinglight/iStock/Getty Images, p. 18 (left); OKRAD/iStock/Getty Images, p. 18 (right); Abdul Razak Latif/Shutterstock, p. 19; Sean Pavone/iStock/Getty Images, pp. 20, 24; powerofforever/E+/Getty Images, p. 21 (top); MasterLu/iStock/Getty Images, p. 22; Sunny Tan/Shutterstock, p. 23; Brent Clark Photography/Moment/Getty Images, p. 26; C. Fredrickson Photography/Moment/Getty Images, p. 30.

Cover: Left_Coast_Photographer/iStock/Getty Images (front); Alfons Hauke/imageBROKER/Getty Images (back).

For Mom, whose love of libraries and fascination with nature inspires my writing journey.

A special thanks to Annie Scott, MS Geology, and Eleanour Snow, PhD Geosciences, for reviewing this book.

Millbrook Press™
An imprint of Lerner Publishing Group, Inc.
241 First Avenue North
Minneapolis, MN 55401 USA

For reading levels and more information, look up this title at www.lernerbooks.com.

Designed by Danielle Carnito

Main body text set in Adrianna Demibold.
Typeface provided by Chank.

Library of Congress Cataloging-in-Publication Data

Names: Kinser, Heather Ferranti, 1967– author.
Title: Nature is a sculptor : weathering and erosion / Heather Ferranti Kinser.
Description: Minneapolis, MN, USA : Millbrook Press, an imprint of Lerner Publishing Group, Inc., [2024] | Includes bibliographical references. | Audience: Ages 5–9 | Audience: Grades K–1 | Summary: "With wind and water, nature shapes our planet in amazing ways. Photos of geological marvels accompany lyrical verse in this exploration of the forces of weathering and erosion"— Provided by publisher.
Identifiers: LCCN 2022049796 (print) | LCCN 2022049797 (ebook) | ISBN 9781728477190 (lib. bdg.) | ISBN 9798765602232 (eb pdf)
Subjects: LCSH: Weathering—Juvenile literature. | Erosion—Juvenile literature.
Classification: LCC QE570 .K54 2024 (print) | LCC QE570 (ebook) | DDC 551.3/02—dc23/eng20230117

LC record available at https://lccn.loc.gov/2022049796
LC ebook record available at https://lccn.loc.gov/2022049797

Manufactured in the United States of America
2-1012684-50712-8/5/2025